I0475458

Copyright © 2011 by Kurt Henke

Dear reader

This book contains inventions and my
great ideas over a period of 45 years.
Some of these inventions I have taken
to Attorneys. After a discussion with
an Attorney I gave several times the
authorization to have a patent search
performed. After completion of the
searches I had another discussion with
the Attorney and I was always
surprised to find out how many patents
in this kind of art were already issued
by the USPTO

Now, that I had my 83rd birthday, it came to my mind, that I can't take all of these inventions and great ideas with me on my last day.

Since I wrote down all these great ideas and inventions, I decided to put them all in a book with the hope that someone will be able to make a good improvement on an invention and develop it into a personal gain. In addition I hope that people like you just by reading this book are inspired enough to come up with an invention of their own.

About the Author

It took me 80 years to build up enough courage to write my first book " The Milk Bench " My Journey from Hindenburg to Jihad - From Sputnik to iPod - from 3 to more than 100 $ a Barrel. Book number 335879. available at Amazon or any book store.

Chapter one

The powerful Ocean waves

During a one day trip, on a Sunday afternoon, I was walking with my family on the Beach in Santa Cruz, CA. I watched the waves rolling in and hitting the Beach. The waves caught my attention and I kept staring at them.

We lived in Los Altos, Ca, and a trip to Santa Cruz came up quite often. Every time we were on the Beach waves would get my attention

After a few month, my wife and I made a week end trip to Carmel by the Sea. As always before, we made a round trip walk on the white sandy Beach and enjoyed the beautiful view. Again I was fascinated by the powerful waves hitting the Beach.

Now, I was interested how much time there was between each wave coming on chore? It was always between 21 and 23 seconds. As soon as we came to the Pebble Beach area my wife would turn to the right and was looking for some Golf balls which some players shot over the cliff and did not recover. We always took an item home we found on the Beach. One day I took all these items: shells, driftwood ,Golf ball and anything of interest and put all these items on a string and hung it from our house in our back yard.

I always lived close to the Oceans. I was born 13 miles inland from the Baltic Sea in East Pomeranian (Pommern) Germany. After age 15 I lived with my ant in Kolberg, two blocks away from the Beach of the Baltic Sea. At the end of WW 2 Poland took over any territory east of the Rivers Oder and Neisse, which included East Pomeranian.

I was very lucky, when I arrived at the Russian Front in St Poelten, a suburb of Vienna, Austria, on May 7th at 2.30 AM when we saw the first German Soldier and asked the question, what is going on here? He said, don't you know the War is over. Here I was standing looking at burned out Tanks and broken Rifles. I did not have a complete uniform and no weapon of any kind. Now, we wanted to get over to the American side. The Demarcation line was the River Ens.. After a rough going and six weeks of time, I arrived at an American POW Camp to get my release papers to get back to Germany. Since I was not 18 years old I was released after 6 weeks, got on the Train in the Direction of

Germany. When on the Train, I realized that I was a displaced Person and could not go home where my Family and friends used to live. It took four years for our Family to reunite. I heard rumors, that many people from my Home area had fled from the Baltic Sea to northern Germany. So, I stayed on the Train. No, not first class - on a Boxcar. I was moving towards the City of Oldenburg near Bremen and close to the North Sea. .I did not find anyone of my Family or friends, but I stayed in Oldenburg for 10 years. I got married in 1952, had one daughter and we immigrated to the USA in 1956. The US allowed 20000 displaced people from Germany to immigrate to the US without going through the regular Immigration Allotment and waiting period.. We were the modern Immigrants flying in on a DC3 to New York and then a 5 day train ride to Tacoma Washington. After 6 month we moved to Seattle and in 1967 we moved to California. At first, to Menlo Park, then Los altos. Now, we are living in beautiful Folsom for 15 year with our children close by.

A weekend trip took us to Moro Bay and one evening, at Dinner time, we were sitting in a Restaurant overlooking the Bay with the Ocean in the distance .Here I watched the waves hitting the Cliffs and moving around. It seemed like the whole Bay was in motion. A question came up in my mind , what would it take to harness this motion power and turn it in to mechanical or electrical energy. We had a nice weekend and went home.

After a few days, I sat down in the evening and put some sketches on paper. There were a lot of sketches laying on and under the table. Finally, I chose one and put my best ideas on it.

You will see this sketch on page

12, and on page 13 you will see the front page of the US Patent issued for a wave motor on Sept. 15. 1903 (It's the oldest one I could find.)

I had put a lot of sketches and drawings in this book. When I was doing my editing on the book I realized that the details of the sketches and drawings were not showing because they were all made on a paper size 8" X 11' and in the reduction to this book size lost the details, so I took all of them out and leave it all up to your imagination.

Chapter two

The Daisy   .

    Several years ago I quit smoking, and now I find the smell of tobacco smoke very offensive. One day I was reading about activated Charcoal. I found a store that carried activated charcoal. Took home a package and was wandering which way I was going to use it to eliminate the smoking smell. When I noticed a three pound Coffee can on the Kitchen counter I got an idea and started to work on it. I made up a screen with holes small enough so the Charcoal would not fall through it and placed it two inches up from the bottom, and punched one quarter holes in to the Coffee can between the screen and the bottom of the can, .for air to enter.. .Now, I went out to buy a table lamp with a high lens and a 40 Watt light bulb, filled the can with activated charcoal, turned on the lamp, and the hot air from the lamp would rise and pull air through the holes on the lower part of the can. Now the smoky air would have to pass through the activated charcoal, rise through the Lamp and enter the room as fresh air. .

My mind was occupied with the

activated charcoal and I was wondering what else I could do with it. Went to a neighborhood Sheet Metal Shop and talked to the owner. I asked him, if he could make me a curtain size sheet metal box? Make 10 at what price. I agreed. We made a deal and in one week I had my ten boxes. Went to South San Francisco and bought 100 pounds of activated charcoal and I found the right 3 speed blower at Granger. I built a container to hold ten pounds of activated charcoal, installed a perforated cover on each end and assembled the whole unit. It was just wonderful. They word got around and I had an inquiry from a Restaurant in Los Altos. I leased 2 units to them. I visited the place many times to see how the units were functioning? The guests made positive comments, and the owner liked very much how the units were working.

I experimented with a "Heppa" Filter 12" x 24" x 12" in size. I completed one unit and later on sold it to a Vintner, who was looking for a filter like this to filter out the Yeast smell at the Winery.in Sanoma County. Later on I built a unit with 50 pounds of activated charcoal and tried it out at various places, including the Computer room at Lockheed. Many people inside the Computer room said it was working very good. The person in charge of this experiment stated at the end: " well", we have our own engineers that are capable of building a unit like this.

Chapter three

Energy Storage Silo

I was visiting a Company and talking to an Engineer who was concerned about energy storage. This statement stayed with me for a while. It was the car Jack that put me on track to come up with an idea.. I was thinking about my wave motor. How about intalling a Floating Pontoon close to the Beach with a long arm and is working like a car jack, and is pumping high pressure fluid in to a row of Silos on land nearby. The head of the Silo may be filled with Sea water. When one Silo is filled to capacity the flow of high pressure fluid is directed to the next Silo. Fill up and raise as many Silos as desired and release them at peak usage times into the Grid.

I was thinking about the Dutch people in Holland. They built Dikes and Locks to keep the Ocean away from their land. How about building Dikes and Locks to keep the Ocean water in, and then let it run out via a turbine and produce mechanical or electrical energy.

.

Chapter four

The wire wheel cover Lock.

I was at a Cadillac Dealership in Los Gatos and I noticed a puzzle lock on a Cadillac sedan. My reaction was: "The are not very good looking" The are on top of the wheel cover and extending outward one inch and a quarter. This was in the seventies, and a lot of wire wheel covers were stolen to get quick cash for a fix on drugs.

I started to think about it to come up with a better looking device. I sat down, early in the morning and made phone calls to Chicago Lock Company to get some information and some ideas.. The last person I talked to stated, that if I wanted something to be fabricated at the Chicago Lock Company they would require a 1000.00 Dollar Deposit and they would call me in six month for a production date. My next step was to call another Lock Company in Wheeling, IL, which was more accommodating. They send me several samples. I went to a Cadillac dealer and bought a single wire wheel cover. Since I had a Cadillac Seville in my garage, I was able to take measurements to come up with the right location of where to place the lock and the bracket where the cam could slide behind. The bracket which I designed would allow the lug nut do go down to its original position .After several trial installations I decided to go ahead and order 600 locks and cams., and found a manufacturer in San Jose, Ca to make 600 brackets of my design with heat treated steel. Now, I went to a printing shop in Sunnyvale, loaded with pictures, hoping to come up with a layout for a brochure. The printer came up with a sample which looked good, and I went ahead and ordered 1000 pieces..

In a month period I got this all put

together and a letter with a brochure
went out by mail to all Cadillac dealers
in the USA. A few days later I got my
first order from Hamburg on the East
Coast. I was so excited, and I wanted to
make for sure the order was real. I
called the dealer and confirmed the
order. From now on the phone just kept
ringing. I had formed a Company three
years ago. Since we were living in Los
Altos, on Catalina Ct., I called the
Company "Catalina Marketing" We
sent out shipments from Buffalo down
to Miami and over to Dallas and San
Diego, up to Denver and over to
Portland. I visited dealerships in the San
Francisco Bay Area and never left a
Cadillac dealership without getting an
order..

Other car makers came up with wire wheel covers such as Oldsmobile, :Lincoln and Chrysler. The distance from the bracket to the Wheel cover was different on each make. I designed a universal cam which was 3 inches long and could be cut to a different length and bend over on location. We used a ½ inch bolt cutter to cut the cam and a vice to bend it over for a proper fit.. For some time business was going very smoothly. I had to make long term commitments to keep the flow of locks coming to satisfy demand.

Dealers were asking us, if we would install the locks for them? We leased a Chevy Van and my son Jeffrey formed his own Company and went to the dealers to install the locks for them. Soon he covered the Bay Area from Concord to Santa Cruz.

It was the time of the next model year when Cadillac announced to all their franchised dealers, that, when a dealer would order a car with wire wheel covers, the dealer would have to buy locks made by Cadillac, or no wire wheel covers. Many dealers, especially myself, were upset because they did not like the locks offered by Cadillac. With the introduction of the next year model our volume of business to the Cadillac dealers went down. The sale of locks to Oldsmobile, Buick and Lincoln did not make up for the volume we had with Cadillac, but we kept going.

Chapter five

The Mercedes Benz Hardtop carrier.

I was at the Mercedes Benz Dealership in Santa Clara, CA and watched workers remove a Hardtop from a 450SL. They stored one on a carrier. I did not like that the weight from the Hardtop rested on the lower back moulding from the Hardtop. After thinking about this for several days, I came up with an idea, the Hardtop had a guide pin on the back and when placed in the proper hole would guide the Hardtop in its proper place, after locked down . I felt this is the piece the Hardtop should rest on. I took a wooden platform, finished it with Linseed oil. Went to a steel outlet and got three pieces of steel cut a curtain length. I would bend it and drill holes in each one myself and painted the pieces black. Now, I ordered a roll of grey fabric, which was suitable to cover the Hardtop when stored.. Packaged all these items including screws and assembly instructions and shipped it to Mercedes Benz dealers after receiving an order. For the dealership in house use we made one that would hold two Hardtops .For customers in the San Francisco Bay Area we made custom style carriers with whatever the customer wanted on it .

Chapter six

The Henke Wind Twin Turbine.

The Henke twin wind turbine is utilizing the Henke wheel patent application # 09/774420 and is mounted on a turret to turn with the direction of the wind. (The Henke wheel is featured in chapter nine.) The twin turbine has a large scoop to harness the wind .The scoop will narrow sharply towards the center of the turbine, thus concentrate the air and build up pressure before entering the turbine. The turbine will drive one generator. One wheel will use a belt and the other one will use a gear drive. A directional control on the top of the scoop will keep the unit facing the direction of the wind..

Chapter seven
River power

A large scoop placed in a river, and mounted on a frame with spikes in front to dig into the river bed and pockets on each side filled with rocks to stabilize the structure. The scoop could be 20 feet wide and 16 feet deep, (or any chosen size) . A large volume of water will enter the scoop and then be squeezed into the narrow path between the two Henke wheels US patent application # 09/774420 for a highly efficient power output. The entrance between the wheels is one foot, six inches for each wheel.. The flow rate of the river and the size of the scoop will determine the output of the small hydro electric plant.

Installation is simple: Dredge the river bottom if needed. Lift total unit with a crane from a truck or barge. Unit should be hanging in a 25 degree angle, pointing downward at the front.

Lower unit until spikes enter the river bed and the unit comes to rest. Now, fill side pockets with rocks.

Chapter eight

White water power and split torque turbine

Objective of Invention: Capture the speed of a white water head and lead it downstream through a system to gain a suitable vertical decline, to gain additional speed and pressure to establish a small hydro electric power station.   The front loader placed into the white water head, and the first few sections of the loader, (depending on the distance) shall be larger than the rest of the loader to insure proper volume for the increased speed and pressure to the split toeque tubine. .

The loader 8 x 4 feet, (or any size suitable for a particular location) provides an even flow of water at speeds up to 8 feet per second to the center of the turbine. It is evenly devided and is turning two wheels. One is turning to the right and the other one is turning to the left, thus providing a split torque for smoother and efficient operation.

In white water areas, the river is running down hill and a desired decline in altitude can be gained in a relatively short distance. The loader shall be placed in a downward angle and be buried in the river bed toward the turbine and extended far enough to obtain the desired decline in altitude.

The front loader shall be about 3 inches below the surface. It will have a hydraulic jack on each side At the beginning a hydraulic pump (12 volt system) will be activated by a mechanical control system having a blunt plate facing the current, which will raise and lower the front loader when the water level changes to seasonal heights. (Where applicable). A limit switch will turn off the hydraulic system at the predetermined highest and lowest point.

The shaft from each wheel will extend at the top and drive a generator . No rocks shall be removed from the installation site, and there shall be no place on the whole structure where water can be collected and remain at he same place.

After a complete installation of the system properly secured by using perforated steel crates, loaded with rocks and attached to the loader

.There are tons of water in the loader to keep it in place. Artificial rocks shall be placed on top of the loader which are not covered by river rock.

A complete functional installation of this system will not block the movement of fish, water sports activities, and the area will look like a natural white water river in its natural

beauty.

Chapter nine

The Henke Wheel

I applied for a patent on this invention. After talking to an attorney, and explaining to him what I was trying to do, I agreed to let him proceed with a patent search. After completion of the search I went to my attorney and discussed the search on hand , and I was always amazed about all the patents that were given out on a particular kind of art. I felt there was a good chance to get a patent for my invention. The patent search did come up with one patent issued, where the inventor used the same idea in a small Fluid drive and I failed to get a patent .

Brief summary of my Invention:

The Henke Wheel is a device designed to maximize the removal of energy from the input forces under pressure such as fluid, air, steam and gas. This is accomplished by removing the force close to the shaft from prior art, and expanding the same area in the same radius on the vane of the Henke wheel and increase the torque 100% over prior art and save energy. The

Henke wheel will not only offer an advantage over prior art in the manufacturing of power equipment , fluid and air motors, it will offer the same advantage in the energy production utilizing free flowing water or hydro electric power equipment.

Chapter ten

Gravity power

I had three inventions, which I wrote down in three separate folders. The first one in May 1981, the second one in 1988 and the third one in 1990. All three had to do with lowering a Cylinder by gravity down to any desired depth. Each cylinder had a moving piston inside. In the first one I used an air compressors to move the piston from the most upper position to the lowest position and convert the cylinder to buoyancy and have it rise to the surface. On the other two I tried to have the Cylinder going down with gravity convert another cylinder to buoyancy by pulling down the piston in another Cylinder and have it rise to the surface and generate mechanical or electrical energy. I was not convinced about the proper functioning until I came home from a weekend trip to Lake Tahoe. The "Light" came on and told me what to do. It is working properly and efficiently and is suitable for commercial energy production. I went ahead and applied for a patent mailed it and now are waiting for a response to get an application number.

Title of Invention:
Buoyancy Conversion Plant.

USPTO Application number 61/210,619

Description:

a 1 ton (or chosen size) Cylinder A parked in the upper docking station, equipped with a reciprocating piston inside the chamber in its most lower position and is releasing the holding mechanism next to each resting cradle via an air cylinder for Cylinder A in a body of water. A mechanical device will be activated by the piston assembly and the existing water pressure will move up the piston in Cylinder A and gravity will pull it vertically downward 500 feet (or any chosen depth) with a force of one ton. An endless chain is connected to the cylinders at the top and the bottom and is running over a sprocket which is mounted at the center of the drive shaft and is turning another drive shaft with a gear mounted at the center which is driving another gear with a ratio of 40 to 1 which is pulling down the piston assembly and bring down the piston in cylinder B, and discharge the water from inside the cylinder A via small outlets totaling 144 square inches. The pressure at 500 feet in a body of water is 226 pound per square inch. The force needed to discharge the water through an outlet of 144 squa4re inches 32544 pounds. The force made available is 40000 pounds or 277 per square inch. The other one half ton is turning a generator with a ratio of 2 to 1.On the gravity stroke the locking pin is forced open by an air cylinder and cylinder B will now rise to the surface with a buoyancy force of one ton and drive a generator, via drive shaft, with a ratio 1 to 1. Before arriving at the docking stations, each cylinder will engage a braking system (an air compressor) and store compressed air in air tank to be used in air cylinders in the locking mechanism. The air pulled in through the air pipe created by the vacuum from the downward moving piston in cylinder B, after docking at the upper docking station, will now be released into an air turbine via an outlet. Each cylinder has an air quick connect function from the top of each cylinder to connect to an air pipe. An air line will provide compressed air from the air tank. Each cylinder has two fins, each fin has a caster which surrounds and is using the steel cables as a guide. After the piston in cylinder B which is resting in the cradle on its docking pins and has moved to its most upper position. The cylinder is decompressed and another cycle may begin.

The direction of motion determines which half of all the sprockets and gears are in the free wheeling or in the performance mode.

The displacement of each cylinder A and B is 64 cubic feet for a one ton cylinder. 32 cubic cubic feet for equilibrium and 32 cubic feet for buoyancy.

The weight of the chain is canceled

out, because the up and down distance is the same

The directional numbers in the description and 4 pages of drawings are

not shown.

Chapter eleven

Oil well fire cap off

At the end of the war "Desert Storm" a lot of oil wells were on fire. The stories in the news paper said, that it would take more than a year to get the fires under .control, I came up with an idea how to put out these fires. I send copies of my invention to the Red Adair Company,

Inc. (Fire and Blowout Specialists) Houston, Texas and to the Embassy of Kuwait (Kuwait Emergency & Recovery Program) Washington, The Oil Ministry from Kuwait send me a reply and stated they were not interested in my invention. All the Oil well fires were out and in control in three months.

Title of Invention: Oil Well Fire Cap off.

Two cranes will lower the twist-on-coupling with a globe valve, and an extension pole on each side, over an existing well fire pipe. The twist on coupling shall be flared on the bottom, for easier mounting with the help of guidance chains. On each side, inside, of the twist-on-coupling there are 5 to 6 bites (or any desired #). Bites are of hardened steel, and the lowest one is diamond tipped . They are mounted in a vertical position. Above and below is an asbestos seal. Ones the twist-on-coupling is slipped over the existing well pipe, (ballast may be used to put down and hold the unit in place) and the oil will now flow through the globe valve and through the extension pipe, Since there is no more oxygen inside the unit, the flame is now above the extension pipe. The flame is now considerably higher of the ground and will allow it to cool.. Once established, the twist-on-coupling will be rammed down 8 to 10 inches, (or any desired distance) and extension booms inserted into the ram receiving arm. Now, the twist-on-coupling shall be turned to go down 2 to 3 inches. Each bit is cutting a groove into the existing well pipe and will hold the twist-on-coupling securely in place. The cutting depth of each bit shall be equal to the depth of a pipe thread with the same diameter. Now, close the globe valve - the fire is out - remove extension pipe and the well is capped off.

PS
Written in March 1991

Chapter twelve

Title of Invention:

Ocean Motion Power

A plurality of bases connected to each other by a boom from below the base and anchored in the Ocean at a chosen place. Each base has a plurality of weights (25 tons each) with a specified displacement and material with a specific gravity so only 10 % of the weights will be above water. Each weight has a boom which is connected and turning on the base, and has a gear connected to the end of the boom which is moving a ram up and down with the motion from the ocean and the waves and the rocking motion of the bases. Each ram has a pipe moving high pressure fluid to a special turbine designed for the task. A return line from the turbine to the fluid reservoir will extend to the ram. Each line at the ram will have a check valve in the return line, opening at 1 psi and holding back a specified psi going in to the ram and the same check valve at the outlet of the ram. The special designed turbine will drive, via transmission, a generator. All hydraulic fluid lines and functions are inside the base for a safe environment. There is a control deck between the platforms and the utility deck above the platforms to provide space for the installation of the most up to date solar and wind power equipment

My Patent application number 09/917.582 had 5 pages 8" x 11" of print and 3 pages of drawings for my Invention: Ocean Motion Power.

A cluster of multiple Bases covering one square mile could produce approx. 12.000 Megawatt plus the energy from the solar panels and wind turbines.

Complete bases can be mass produced and towed to almost any place in the

world

Chapter thirteen

The Henke Turbine

The Henke turbine is a twin turbine were forces such as fluid, air, steam and gas enter via a manifold at the center of the Henke turbine between the Henke wheel thus dividing the entering forces to turn one wheel to the right and the other wheel to the left, to help stabilize the whole structure by not having the torque twist associated with a single wheel. The Henke Turbine will operate in the horizontal, vertical and parallel position.

To get the final drive going in one direction only, since one wheel is turning right and the other one to the left, have one unit use a gear drive and the other one a belt or chain drive.

Chapter fourteen

The Energy Ramp

Capture the energy from the weight of of a moving vehicle going down a grade, or whenever it is coasting. How about the train going up to 7500 feet and than come down each side of the Sierra or Rocky Mountains. To harness the enormous breaking energy required to slow down the total weight of a train to bring it down the Mountain safe .At Truck stops, Toll Plazas.

A multiple number of heads from air cylinders could be installed in the road bed with the heads extending a suitable distance above the surface, for the vehicles to drive over and a group of air cylinders are merging to push a high pressure fluid cylinder which will discharge the pressurized fluid via a check valve into a fluid motor to drive a generator to produce electrical or mechanical energy .On slower moving vehicles a system without the air cylinders would be functioning just fine.

A good system for the Railroad could be a row of fluid cylinders on the inside of the track close to the rails, with leavers attached to the cylinder rod and close enough to the rail for the inside of the train wheel to run over and to push down the Cylinder to create high fluid pressure for a fluid motor to produce electrical or mechanical energy.

Another system for the Rail Road coming down a grade could be were the moving train would engage a cable and harness the energy from a moving train: just like the San Francisco Cable cars engage a cable and be pulled up a grade.

Chapter fifteen

Aquatic Separator and River flow control

My invention is novel. It separates a river, as close as possible, at the estuary (rivers end) from the ocean and allows pumping of fresh water in to a distribution system at the daily c.f.s.* flow rate, or hold back the river flow when the daily c.f.s. flow rate is very low, to improve wet land conditions and the environment for marine life upstream, or open a gate and let some fresh water enter the ocean and some to be pumped in to a distribution system. The aquatic separator will have on gate open all the time for migrating fish to leave and enter the river .Second choice: a tunnel below the aquatic separator or a fish ladder to accomplish the same. Provide Locks for the aquatic separator, where the channel is deep enough for vessels to leave for the ocean and return to the river.

** It is amazing that of the worlds water 97% is in the oceans and 2% in the form of ice and 1% of the worlds

water is going round the water cycle at any one time. The energy from the sun will evaporate water from the oceans, the rivers, lakes and plants which will rise in to the atmosphere, return as rain and fill the rivers, lakes and return some of it back to the oceans. The water volume in the world is constant and it is not necessary to return all the fresh water to the oceans in order to continue the water cycle. Each year the rivers of the world deposit eight thousand million tons of sediment in to the oceans.. **
(Author unknown)

The discharge of sediment can be stopped with the aquatic separator and the sediment could be harvested for improvement of agricultural land.

* cubic feet per second.

Chapter sixteen

Solar oven

I was thinking about a solar oven in my back yard . A circular structure about 10 feet high, 6 feet wide at the bottom and a 2.5 feet opening at the top and a Henke twin turbine mounted on the top..

A structure of sheet metal supported with structural beams and a lot of heat absorbing tubes inside. An 8 foot wide and 1 foot high circular structure covered with black plastic panels and openings for air to enter in. The sun will heat the whole structure, the hot air will rise and drive the twin turbine.

I turned on my computer and typed in solar oven. I found out they are building one in Australia. A huge structure. It was about 3 years ago and I do not remember the measurements, I believe it was about 200 feet high and 20 feet wide at the top. They were expecting to produce enough energy to supply thousands of homes.

I think it is a good idea for our local energy supplier SMUD to convert the two stripped down towers left over from their Nuclear power plant which the California voters wanted shut down.

I believe the 2 towers are very suitable for a conversion to solar ovens and be very productive and good for the environment

Chapter seventeen

wind power

Wind turbines are used all over the world. The power grid leaving our locale Folsom Dam is almost running past my back yard. I see it every day. One day I was asking myself, why don't they utilize these high, strong structures and mount suitable wind turbines on the side and on top of the towers and charge the energy right in to the grid. Especially in rural areas. No power lines needed to feed the energy in to the grid.

Chapter eighteen

Fresh water for California.

It is in the daily <u>news:</u> We need more water. We need more water storage. We need to build another dam. When we came to California in 1967 there were about 19 million People in California. Today there more than twice the amount. How many will there be here in 10 years from now? The rain and the snow fall; did not double and will not increase in the future.

Our Folsom dam is controlled by the
Federal Government and the release
water when they think it is necessary to
Keep the fresh water flow strong enough
to keep the salt water in the bay and
from entering the Delta.

I am sure a lot of people are thinking
about and are concerned about the water
supply in the future. Why not build a wall
at the narrowest point in Suisun Bay and
keep the salt water on one side and all the
fresh water on the other side in the Delta.
A separating wall would not be as costly
as a Dam since the pressure would be, for
the most part equal on each side

Chapter nineteen

High Heel Hugger

Patent application number

The Heel Hugger is a device that slips over a high heel to provide flooring protection and eliminate damage, especially on softer floor covering material, such as wood and linoleum. The ground layer shall not have any sharp edges and be at least one square inch in size. The second layer shall be the same material as the ground layer, except it has a 17 mm hole at he center for the heel to enter and to rest in. The outside cover shall be of plastic with a cone shaped foam through which the heel can enter and be held in place. An elastic strap is attached to the heel hugger and can be pulled up to the highest position on the heel and be attached to velcro to hold the heel hugger in place while in use .The heel hugger is designed for heel tips in size from 9 to 15 mm in diameter.

The directional numbers in the description and two pages of drawings are not shown.

Chapter twenty

Exhaust Arrester

I was in the Automotive Business for 42 years. When I was on a showroom floor filled with cars and a car had to be removed or put on the showroom floor, it had to be started and moved and the exhaust smell would be there for an extended period of time. This got my attention and I was determined to do something about it. After some time I found a none-toxic chemical sorbant carrying the Underwriters Laboratories

Class 1 rating: called Purafill. It is used by major industrial plants throughout the world.

Now, the question came up, how can I use this chemical to filter out the exhaust smell and even carbon monoxide from a car, when started in a closed environment.. I again went to the same sheet metal shop and had him make up a box that would hold 15 pounds of Purafill were the exhaust fumes could pass through and a place were I could attach 2 receiving hoses to handle a car with one or two exhaust pipes. At the end of each hose there was a pipe with a check valve. A rubber bag with a hand pump and pressure gauge (like the Dr uses to check your blood pressure) is attached to the pipe. The rubber bag is attached to the pipe and the pipe is inserted in to the exhaust pipe. After pumping some air in to the bag up to a curtain pressure, a seal is made between the exhaust pipe and the end pipe of the exhaust arrester. Now, all the exhaust fumes are flowing towards the exhaust arrester and are guided through an air filter to remove moisture from the fumes and then slow down many times and move trough the Purafill which will remove 95 % of the odorous gases and a large portion of carbon monoxide. A 100% removal of carbon monoxide is not attainable. The unit could just be set down in the trunk of the car, then attach the hoses to the exhaust pipe or pipes and start the car (do not accelerate) and move the car with the lowest rpm possible to prolong the life of Purafill. Now remove the receiving hoses, store unit on a moveable platform, cover it with a neutral color vinyl cover and store it on the showroom floor. With average usage at a dealership the Purafill should last 90 days or as soon as you smell the exhaust fumes

This was 1985 and I had planed to sell or lease the unit to dealerships across the country, but I did not have the proper funds to market the product and dropped the idea..